How to Read
a Chicken's Mind

How to Read a Chicken's Mind

Understanding How Chickens Learn,
Perceive People, Express Emotions,
and Pass Down Knowledge

Melissa Caughey

Storey Publishing

The mission of Storey Publishing is to serve our customers by
publishing practical information that encourages
personal independence in harmony with the environment.

EDITED BY Lisa H. Hiley
ART DIRECTION AND BOOK DESIGN BY
 Alethea Morrison
TEXT PRODUCTION BY Jennifer Jepson
 Smith

COVER PHOTOGRAPHY BY © Elena_
 Goncharova/Shutterstock.com, back;
 © Piven Aleksandr/Shutterstock.com,
 front; Terri Calla Photography, back
 (author)
INTERIOR PHOTOGRAPHY CREDITS ON
 page 128
ILLUSTRATIONS BY © Katie O'Shea Design

Storey books may be purchased in bulk for
business, educational, or promotional use.
Special editions or book excerpts can also be
created to specification. For details, please
contact your local bookseller or the Hachette
Book Group Special Markets Department at
special.markets@hbgusa.com.

Storey Publishing
210 MASS MoCA Way
North Adams, MA 01247
storey.com

Storey Publishing is an imprint of Workman
Publishing, a division of Hachette Book
Group, Inc., 1290 Avenue of the Americas,
New York, NY 10104. The Storey Publishing
name and logo are registered trademarks of
Hachette Book Group, Inc.

ISBNs: 978-1-63586-868-5 (paperback);
978-1-63586-869-2 (ebook)

Printed in China by Toppan Leefung Printing
Ltd. on paper from responsible sources
10 9 8 7 6 5 4 3 2 1

Library of Congress Cataloging-in-
 Publication Data on file

DEDICATED TO ALL MOTHER HENS—
including my own, Joyce.
They are the ultimate mind readers.

CONTENTS

PREFACE

"This is interesting—very interesting—something quite new. Give me the Birds' A.B.C. first—slowly now."

So that was the way the Doctor came to know that animals had a language of their own and could talk to one another.

—HUGH LOFTING
The Story of Doctor Dolittle

When I wrote my book *How to Speak Chicken,* I put myself out there as a self-proclaimed "chicken lady." As I was writing, I wondered if I was alone in my thoughts and observations about my flock. I was excited to share what I'd learned so far. Now I smile when I get emails from people asking me to assist in deciphering what their chickens are saying. It's so fun to have those aha moments with fellow flock keepers!

But almost as soon as that book was published, I started wondering, "What if I've only scratched the surface?" More and more, people are curious about the minds and inner lives of all animals. Newsstands feature periodicals proclaiming the brainpower of dogs and cats and many others. Many books have been published about how animals think and feel and experience the world. I love to follow the science and am always learning more, especially about our feathered friends. It turns out that mammals and birds are quite similar in terms of sleep processes, metabolic function, socialization, and brain size relative to body size.

I believe we are living in a renaissance of understanding animals' intelligence and capabilities. For centuries, humans viewed animals as inferior, considering them to be unthinking and

unfeeling. Animals had no purpose other than to serve people's needs for food or labor, or to amaze and amuse us with their beauty or behavior. But as our understanding of and regard for animals grows, we are realizing we don't have the full picture. All animals, including close-to-my-heart chickens, have more to share. When we find common threads that weave our lives together, our relationships with animals improve, and we can connect with them on a deeper level, more so than some folks would admit.

So here I am, pecking away at the keyboard again, sharing oodles of chicken facts and personal observations, along with some science, that might make you look at chickens in a whole new way, whether you are already a dedicated chicken keeper or have just been interested in them from afar. Chickens are marvelous creatures, underappreciated, downplayed, and trivialized. I hope this book brings you many aha moments of your own and teaches you a bit about how to read a chicken's mind.

Melissa

1

THE UNIVERSAL BOND

HUMANS AND CHICKENS

Since the earliest days of civilization, chickens have had a special relationship with humans. For centuries, they have not only provided a vital and affordable source of food and income but also served as companions, friends, and even family members. While it's hard to pinpoint exactly how many billions of chickens there are in the world today, it's safe to say there are more of them than any other kind of bird!

For many backyard chicken keepers, tending to their flock has become much more than a hobby. With chickens available in such a wide variety of breeds, with different egg colors and characteristics, not to mention their individual personalities, it's easy to see why so many of us just can't help adding one or two more hens. We're all familiar with "chicken math," where no one can figure out how the original 5 chicks turned into a flock of 15 or 20!

The Domestication of Chickens

The first evidence of human-chicken interaction was discovered in China and dates to 5400 BCE, but scientists have confirmed that true domestication did not occur until 3600 BCE in Thailand. German zooarcheologist Joris Peters and his colleagues theorized that chickens became domesticated through human cultivation of rice and other grains.

They speculate that once people began growing and storing these crops in Southeast Asia, jungle fowl came out of the forests to exploit these new sources of plentiful food. Over time, chickens started living near the open fields and are believed to have assisted farmers through their keen ability to sense intruders and alert the village. Chicken "watchdogs" benefited the villagers in exchange for a bit of food, and the long process of domestication began as this symbiotic relationship flourished.

Chickens appeared in Europe about 1,000 years after they were domesticated in Asia, based on bones found in Italy that date to the first millennium CE.

Interestingly, the researchers also hypothesize that people didn't always eat chickens. Chicken bones discovered in the ancient graves of farmers and their families in Thailand may indicate that chickens were an important part of people's lives, perhaps even kept as pets.

Another possibility is that chickens were believed to be spiritual entities that helped guide one safely to the afterlife. I had to smile when I read this. Perhaps it confirms my belief that chickens go to heaven.

Modern chicken breeds are descended from jungle fowl that were domesticated by farmers in Southeast Asia thousands of years ago.

Red jungle fowl

How Chickens Are Still a Lot like Jungle Fowl

✦ Even in their coop, chickens prefer to roost high up, where they are safer from predators than they would be on the ground.

✦ Chickens are omnivorous foragers, supplementing their commercial diet with insects, snails, slugs, worms, frogs, snakes, and small mammals.

✦ The flock's hierarchy of a dominant male and/or female and subordinate flock members with defined roles imposes order and streamlines communication to optimize food gathering and response to threats.

✦ Roosters serve as the ultimate flock protectors, fiercely defending their charges from threats, including other roosters.

✦ Roosters freely mate with all willing females, but hens have a say in who they accept.

✦ Hens remain fertile for four days after mating, increasing the odds that their eggs will develop chicks.

The red jungle fowl (*Gallus gallus*) is considered the primary ancestor of the domestic chicken. It ranges across much of Southeast Asia and parts of South Asia. The other three species of wild jungle fowl—gray, green, and Sri Lankan—also contributed genetic material.

How People Perceive Chickens

How often have you heard the term *birdbrain*? Even though birds traditionally have been perceived as dumb, decades of research show that many birds have the same degree of intellectual capacity, social behaviors, and emotional depth as mammals. But many people still think that chickens fall on the lower end of the intelligence spectrum. I suspect that this is because, even though chickens are the most common domesticated animal in the world, many people have never met any chickens in person, let alone gotten to know them.

Let's face it, chickens are primarily valued as a source of food. Most people view chickens as a commodity rather than as individuals with unique personalities. Because of this, poultry research has focused on technical ways to generate more eggs and improve the quality and size of meat birds for harvest. But in the past few years, an interesting thing has happened: Farmers have realized that chickens kept in calm conditions with sufficient space lay more eggs and exhibit fewer aggressive behaviors than those kept in crowded, stressful conditions.

If you take the time to observe and interact with chickens, you'll soon learn that they are all individual characters.

Even though the value of these birds is mostly calculated in economic terms, I believe that view often changes once people experience life with chickens. The ability of chickens to experience emotions makes them relatable to us on some level. We can form connections because we share emotional ties with them. That's why they make great pets!

Getting to Know You

An experiment by an Australian research team showed how people's perceptions change when they get to know chickens up close. Ninety-four chicken-naïve college students completed a questionnaire to determine what preconceived notions they had about the intelligence level and other attributes of chickens. Not surprisingly, most of the students had a low opinion of chickens and thought of them as slow learners. Most students did believe that chickens could feel pain and hunger and could experience fear, but not more complex emotions.

Following the initial questionnaire, the students were charged with using clicker training to teach chickens to peck a red circle and avoid a green circle to earn food rewards. After eight weeks, the students noted that the chickens learned at different rates and exhibited unique personalities. Not surprisingly, when they completed another questionnaire after the training sessions, the students' attitudes toward chickens had changed greatly. Initially, only 7 percent had a high opinion of chickens; at the end of the study, that number rose to 61 percent. Attitudes about chickens' intelligence, personalities, and ability to experience frustration, boredom, and happiness also improved!

How Chickens Perceive People

When you first meet chicks, they will usually be skittish around you, doing their best to avoid being picked up and, when you do catch them, trying to wriggle out of your hands. It's natural for chicks to want to escape from something as large and as foreign as you are, especially when they are removed from their tiny peeping flock. You would be scared, too!

However, with time, patience, and observation, you can begin to form a relationship. The chicks watch you dole out food, fill the waterer, and tidy their living space. They learn that they don't need to fear you when you are performing those duties. They quickly realize that you are a source of treats as well! Over time, they become accustomed to your presence and your routines, and they begin to recognize you and can tell you apart from other people. When we have visitors that the chickens have not met before, they are skittish and tend to avoid the strangers until they develop trust and get to know them.

I know that my chickens recognize my voice without actually seeing me. In the summer, I shout to them from the kitchen window to let them know I am coming. They can't see me, but when they hear my voice, they begin jumping and chatting with excitement. But I have often wondered how my chickens recognize me without hearing my voice.

I am sure that my chickens have a general idea that I am a human, but how do they know that I am different from other humans? I have noticed that when I am wearing a heavy hooded coat, my chickens do not know it is me. They act nervous and

won't come to greet me until I begin speaking or take down my hood. Do they use my hairstyle or hair color to identify me, much like they identify one another by the shape of the combs on their heads?

Do My Chickens Love Me?

Some chickens clearly enjoy the company of people and actively seek opportunities to interact. Other chickens are more aloof and prefer interacting with their flockmates. Many of us wonder how we can know what our chickens think about us. We want to know, do they love us?

There are a few ways you can tell that a chicken feels affection toward you. A hen may choose to settle down near you while you are working in the garden or having dinner on the patio. If a chicken sits at your feet and is content to stay put, you can be sure she feels safe and wants to be near you. An outgoing chicken might come to you and weave between your legs, almost like a cat, to show affection.

When chickens look up and connect with your gaze, they are letting you know that they want to engage. Some chickens enjoy spending time in your lap and might even preen you or put their head over your shoulder and into the crook of your neck by your ear. Then they will snuggle in with their eyes closed. You can feel little puffs of air from their nostrils. I like to think of it as the way they hug us back.

●

"Chicken scratch," the grains that chicken keepers
toss out to their flocks, also refers to handwriting that is
so poor it looks like the marks in the dirt made by
chickens scratching for food.

●

You can encourage and nurture those feelings of affection by taking time to just hang out with your flock and pay attention to them. Talk to them, move gently around them, and build up their trust by offering them treats from your hands instead of always tossing scratch onto the ground. Any relationship worth having takes time and effort.

Chickens as Healers

By now, it is widely understood that interacting with animals can decrease stress, relieve anxiety, improve heart health, lower blood pressure, reduce loneliness, boost mood, and provide feelings of social support for people of all ages. When people think of therapy animals, dogs are usually the first to come to mind, but many animals can provide that healing sense of connection. For example, interacting with classroom guinea pigs or other pets can help children become calmer and decrease their anxiety levels. Studies have found that kids who read to animals have better social skills, are more cooperative, volunteer more, and share more than others.

The ability to connect with chickens spans all age groups and all health conditions, so it should come as no surprise that therapy chickens have been used in schools and in programs for kids with autism and other neurological or mental health challenges. Likewise, they have found their way into group homes for those with physical and cognitive impairments and in nursing and retirement homes.

For example, several dozen memory care centers in the United States and Great Britain welcome regular visits from therapy chickens. Some have even added chicken coops to their grounds, and residents enjoy caring for and interacting with the flock. No matter the setting, chickens can provide support for those who are experiencing loneliness, isolation, anxiety, or depression and can serve to improve feelings of being loved and needed.

Scientists have studied the effects of chickens in these settings, especially in facilities where patients suffer from dementia or certain psychiatric illnesses. They've noticed an improvement in overall condition, improved behaviors, and in some cases even a reduction in the need for medication.

In addition, parents of children with autism have reported that keeping chickens has made their kids more physically active, improved their social skills, and encouraged verbalization. (Who doesn't love to talk about their chickens?) Chickens and other animals can sometimes do what other types of therapy can't by giving children positive, nonjudgmental ways to connect with another being.

Developing Trust

Developing trust with chickens, as with people, takes time, energy, patience, and kindness. Trust does not happen overnight, but when you put in the effort, you can create the atmosphere for a friendly and happy flock to form.

When attempting to gain the trust of a chicken, you'll need to remember a few things.

- ✦ Patience is key. The process can take a couple of months, so don't give up.

- ✦ Interact with your chickens in a safe, quiet place, such as adjacent to their brooder, in the chicken run, or in your yard.

- ✦ Talk to them! You'll be amazed at how much they respond to your voice.

- ✦ Move slowly and deliberately to avoid triggering a flight response in your birds.

- ✦ Spend time with your flock when you feel calm, quiet, and at peace. You want to emit tranquil energy.

- ✦ Allow your interactions to be on their terms, not yours. Let them approach you instead of the other way around.

- ✦ Handing out treats to your chickens is a good way to build a connection and increase their confidence.

It is easiest to build close relationships if you start with day-old chicks. This allows you to handle them frequently and establish your role as a caretaker. You can even try to mimic their happy peeping noises. Even though chicks are naturally skittish, they learn quickly.

With patient and consistent handling, even older chickens soon recognize those who interact with them and learn not to fear them. Chickens remember and will avoid people who have been mean to them. They can also recall those who are kind. This trait may help bridge your acceptance into the lives of new chickens in your flock. When chickens trust you, they will seek you out and approach you. Instead of shying away from you when you do the chicken chores, they will come and investigate. They will eat from your hands and may even hop into your lap for a visit.

Imprinting

Newly hatched chicks are innately wired to bond with the first moving object they see, a process called imprinting. In nature, the bond typically develops between the chicks and the parent birds, ensuring that the young will follow the parents closely to learn how to find food, take shelter, and avoid danger. Wild birds reared in zoos or animal sanctuaries are often fed by humans wearing hand puppets that look like adult birds. This prevents them from imprinting on their human caregivers and maximizes the chance that they can be released back to the wild.

If you raise day-old mail-order chicks, there is a chance that they will imprint on you or one of your family members. The imprinting process occurs over several hours of exposure to the same individual. As you care for the chicks, they associate you with the things they need to survive.

2

FLOCK TALK

HOW CHICKENS COMMUNICATE

A Silkie or Buff Orpington may not look like a jungle fowl, but it turns out that chickens retain many instincts and behaviors of their wild ancestors. American zoologist Nicholas Collias and Elsie Collias, his research partner and wife, worked for nearly 50 years to decipher the language of jungle fowl in Malaysia. Their fieldwork, plus 6 years studying the vocalizations of a flock of jungle fowl at the San Diego Zoo, enabled them to decode more than 20 different vocalizations, including attraction calls, alarm cries, and threat sounds.

Thanks to their body of work, we know that our backyard chickens have a lot in common with the wild birds that still roam the forests of South and Southeast Asia.

I Recognize That Voice!

You know when someone calls to you across a crowded room and you know who it is just from their voice? Or when you watch an animated movie and immediately identify the actor voicing a particular character? Vocal uniqueness is sort of like a fingerprint. Your vocal pattern is determined by the vibration of air passing over your vocal cords and being shaped by your throat and facial structure. The pitch of your voice is determined by the tension and length of your vocal cords.

Chickens do not have vocal cords, but like nearly all birds, they have a unique vocal structure called a syrinx. A bunch of chatting chickens may all sound the same to most people, but just as many backyard chicken keepers can tell you who lays which eggs, I can identify several of my hens just from hearing them. Especially Lucy.

I wonder if my girls would be able to communicate with a flock of jungle fowl—perhaps they speak in different accents!

Every morning, I toss some scratch into the run for the girls while I clean out the coop. Without fail, I hear Lucy calling, followed by the click of her nails as she climbs the ramp back into the coop. Her voice is so deep, low, and raspy that I wish I'd named her Lauren Bawkall! When she reaches me, she squats for a quick pet.

Lucy's distinctive voice sets her apart from the others, but I can also tell when it's Fluffy or Cuddles coming to carry out the work of visiting me and inspecting my morning chores. Fluffy is soft-spoken, with a whispery voice that is timid and song-like. Cuddles has a stronger voice, more confident and deeper in tone. Do you know any hens with distinctive voices?

Chickens Can Understand Other Species

Many people learn a second or even third language, either in school or from their family. While it can be challenging to become fluent in a new language, it is not difficult to pick up and use a few words and phrases. If you have traveled to a place where you don't speak the language, you realize that body language, facial expressions, and gestures are often universal. It turns out that animals can also learn bits and pieces of "vocabulary" used by other species.

For instance, it is well known that predator calls are understood among many different species of birds and mammals. These alarm calls can carry important information, such as the type of predator, its proximity, and whether it is in motion or standing still. One example is the prairie dog, which gives different calls to warn of snakes and hawks. Burrowing owls, which live in abandoned prairie dog burrows and are preyed on by the same hunters, benefit from the watchfulness of their neighbors and respond differently when they hear the different alarm calls, checking their surroundings for a snake or looking up to spot a hawk.

I witnessed this phenomenon in person when a flock of crows began to frequent the treetops in my yard. Crows are highly intelligent and very vocal. They are always alert to their environment, keeping an eye out for potential predators. A crow that spots a threat lets out a "Haaaa, ha, haa, ha, ha, haa, ha," with each "ha" becoming louder, more staccato, and more purposeful. The other crows join in, as if to say, "Yes, I see it, too—look out, everyone!"

After a while, I realized that my chickens had learned this call, though of course it isn't a sound they make themselves. They trust the crows, and when they hear them calling, they stop what they are doing and assess the situation, looking around to see what is causing the ruckus, then seek out safety as needed.

While the flock forages, the sentinel hen monitors the surroundings and listens for warning calls from other birds.

Chickens Can Understand Us, Too

What about chickens learning English (or French or German)? I'm convinced that, just like pet dogs and cats, chickens can understand some words and phrases in human languages. When I talk to my hens, they cock their heads from side to side while gazing up at me. They listen and seem to be trying to understand what I am saying. With time and practice, chickens will learn words that are significant to them, usually ones that are connected to a reward. Most chickens are quite motivated by treats!

For example, my whole flock comes running when they hear me call out "Giiirrrlllss!" They know that something good usually follows that summons. But when they are in the garden or another space where they shouldn't be, they understand that my quick "Shooot" means they need to move on to different scratching grounds.

Over the course of the years, some of my chickens have learned to recognize their own names and will come when called, just like a dog or cat. It's not something that I set out to teach them, but I've noticed that some hens who seek out my company and enjoy one-on-one time eventually recognize their individual names and will come running when I call them. I think it's because I use their names frequently while talking to them. This does take time and is not something that they learn right away.

One year, a flock of rogue guinea hens moved onto our property. They quickly learned the special call I used to summon them from their free-ranging adventures. "Here, chick-chicks" would bring them waddling as fast as they could to come visit with me and get some treats.

Chickens are quite capable of making their opinions known!

Chicken Chatter

Chickens are continually communicating: They relay information about their environment, chat about life in the flock, and use their language to build one-on-one relationships. Chicken talk, like human speech, greatly depends on tones and pitches, both high and low. No one wants to listen to a droning monotone, not even a chicken! When we converse, our voices rise and fall. You can tell if someone is happy, upset, sad, or excited by listening to the tone of their voice. It's the same in music, where notes climbing up a scale evoke rising emotions, while falling notes evoke sadness or a calming of tension.

In addition to using rising and falling inflection, both humans and chickens communicate through the rate at which they vocalize. Depending on what we wish to stress, we might speak slower or faster than normal, which helps the listener ascertain our meaning. A chick taken from the brooder will peep in distress. Her calls become more frequent the longer she is away from the brooder and the more fearful she becomes.

Volume matters, too. Like us, chickens sometimes speak in whispers and at other times screech so loudly that you cannot comprehend how such a powerful sound can come from such a small animal. Soft, slow sounds indicate contentment and relaxation. Loud, shrill squawks are associated with fear and pain or are intended to grab the rest of the flock's attention. Normal chicken speak is like quiet chatter. It is held at normal conversational volume, just as you and I would hold a dialog.

Using Fowl Language

I started communicating with chickens without even thinking about it. Driving home from the post office with our first box of day-old chicks, I was enraptured by the variety of cheeps and peeps coming from my tiny passengers. I wondered if the radio was bothering them, so I turned it off. I was worried that they were frightened, and it came naturally to try to console them with soft strings of tender words, like a mother would say to her children. "It's okay. You are going to be all right. Just a bit longer and you'll be home. I can't wait to meet you, little ones!"

As the days passed, I kept talking to the chicks, and I found myself mimicking some of their little calls. I didn't know what they were saying, but I wanted to be a part of their world. I wanted to do everything possible to be a good chicken mama. Over many months of immersing myself in their world, observing their behavior, and repeating the sounds I was hearing, the mystery of "chicken speak" began to unravel before me.

Building an Advanced Vocabulary

At first I was able to decipher only the more obvious chicken calls. Here's a quick list of the basics of chicken language (I cover these in more detail in *How to Speak Chicken*). Some of these sounds are made by both hens and roosters; others are specific to sex.

+ Buh-dup (Hello.)

+ Doh-doh-doh (Good night.)

+ Bwah, bwah, bwah, bwah (I'm about to lay an egg.)

+ Buh-gaw-gawk, buh-gaw-gawk, buh-gaw-gawk (I laid an egg! Hurray for me!)

+ Grrrrr, buk, buk, buk, buk, buk, buk, BUKGAW (Look out, danger alert!)

After more than a decade of keeping chickens, I'm still learning what these fascinating birds have to say. I'm excited to share more of their language with you.

Acknowledgment

SQUAAAWKK

This dissonant call, almost a yell, is often made in response to being pecked by another hen who is asserting her place in the pecking order. It may also be heard during flock discourse and when expressing disapproval of male attention. It's a dramatic sound, with moderate to loud volume. Depending on the circumstances, I think of it as, "Okay, I'll leave you alone" or "I don't agree" or "Hey, I'm mad."

A broody hen will tell you in no uncertain terms that she'll peck if you come too close to her nest!

Warning
SSSSSSSSSSS

Sounding like a cat or snake, a broody hen warns intruders to stay away.

Surprise
Uh uh uh uh uh uht

This staccato sound, with the emphasis on the final syllable, is heard when chickens are startled by something unexpected or when a flock member is missing. I like to think the chickens are saying, "Oh, hey, hey, hey. Hey!" or "Wait, is everyone here?"

Frustration 1
Raaahhhhhnnnnnnnnn, awwt, awwwwt, aawwwwt

This sound, described as the gakel call by researchers, starts off midrange and features a long, drawn-out whine followed by shorter vocalizations that dip in pitch, return to the normal pitch, and finally rise in pitch. Given at normal volume, this sound expresses frustration, as when a chicken sees food but is not able to reach it.

Frustration 2
Uhhhhhhhhh

This quiet, high-pitched whine also expresses frustration. It can be heard alone and sometimes in response to hearing the gakel call from another chicken. When chickens use this as a response call, it is an endorsement of the other hen's feelings of frustration, as if to say, "Yes, I agree with you, I'm frustrated, too."

Anticipation

Bwh, Bwh, Bwh, Bwh

Although similar to the "bwah, bwah, bwah, bwah" call
that signifies that a hen is about to lay, the syllables of the
anticipation call are shorter and the tone is much less dramatic.
In response to this call, other hens sort of chatter along with
quiet, higher-pitched trills, almost as if they are saying that they
are just as excited as the hen who is speaking up. You might
notice this call while you refill food dishes, clean the coop, add
wood ash to the dust bath, or do other things that make chickens
happy. I also hear this call when a multiparty dust bath is
underway.

Happiness

Bup ba bup ba bup BUP

Listen for this excited call after a hen emerges from dust bathing,
as she greets the sight of a handful of treats in your hand, or as
the flock frolics around the yard and garden.

Contentment

Prrrrrrrrrrrrrr

This is the same sound made by baby chicks but with grown-up
voices. Both hens and roosters purr, though with roosters, it
sounds more like growling! They make this sound when they
are relaxed, sleeping, or snuggled in your arms. You may hear
it during dust bathing. Sometimes purring is accompanied by
happy little clucks.

Chickens give two types of pecks to one another—
one gentle, like a tap to act as a reminder, and the other
aggressive, mean, deliberate, and perhaps a bit painful. The
receiver knows the difference and may give the surprise
call—"Uh uh uh uh uh uht"—in response.

Pay attention!

Duk, Duk, Duk, hlllllllllllll

A hen makes this sound to call her little ones to safety or tell them to stay close when they are exploring. The last part starts with a breathy sound and moves into a trill, as if you are rolling your *R*s but with *L*s instead. If she fears danger, Mama will use this call, then spread out her wings as the chicks take cover underneath.

Time to Eat!

DooK, dooK, dooK, dooK du dooK

This call is quiet, fast, and low-pitched, with an emphasis on the *K* sound. A mama hen makes this call to gather her chicks around her to share food and water. Sometimes she combines the call with tapping her beak on a hard surface, like a feeder or waterer.

When a Rooster Crows

Ur, ur, ur, UR-URRR

The crow of a rooster is often synonymous with living on a farm. Although famous for crowing at dawn, roosters crow whenever they want to—sometimes even in the middle of the night. They crow for many reasons, including to announce their presence, assert themselves among other rooters, express pride, and alert others in the flock to perceived threats. Cleverly, they will also crow to mislead a predator about the flock's location, drawing attention to themselves while the rest of the flock takes cover.

There are a few unique calls that roosters make throughout the day during their activities and interactions with the rest of the flock. Here are some vocalizations you might notice.

Courtship Call

Buh, buh, buh buh, huuuuuh

This call—a series of staccato clucks followed by a moan that rises in pitch—accompanies tidbitting, the dance done by a courting rooster as he bobs his head, lowers his wings, and circles around a hen, picking up bits of food and dropping them for her. It is a low-pitched sound, quiet and subtle, as if he is speaking only to the interested hen.

Food Call

Uh oop, uh oop, uh opp, uh, uh

Roosters have a special call to let their girls know that they have food to share. Although a rooster will pick up and drop the food to solicit the attention of his hens, he isn't dancing when he makes this sound; it is just a vocalization to alert the hens to a tasty morsel. The pitch of this call varies up and down with quick staccato notes. It is a sweet, gentle, soft sound meant to entice the hens to come and see what the rooster has found, whether it's a bug, a mealworm, or just some scratch.

Keeping the Peace

DuK, duK, duK

The rooster makes this sound—quick little clucks with the emphasis on the *K*—in response to hens' disagreements and squabbles. It is very similar to the sound that a mother hen makes to her chicks when she is calling them for food and water.

●

Compared with beta, or less dominant, roosters, alpha roosters are quicker and more agile, have larger combs, and are more aggressive. After being startled, they remain vigilant for long durations.

●

War Cry

Urrrrrrrrrrrr, Urrrrrrrrrrrr

Before a rooster attacks a perceived threat, be it another rooster,
a predator, or a person, he gives an alarm call that sends the hens
seeking cover. This war cry sounds like a high-pitched scream.
Perhaps he is saying, "I'm going in!"

Chickens Choosing Names

Years ago, with the help of Sy Montgomery, an animal expert and my friend, I realized that one of my chickens, Tilly, had a special name for me. (In case you're wondering, it is "Bup, bup, bup, baaahhh," starting low and rising almost an octave.) It was an exciting discovery, and it got me wondering if chickens also had names for one another. From reading about animal communication and observing my own flock, I've come to believe that chickens, like dolphins and some other animals, use individual names for members of their flock.

Scientists call these "signature contact calls." We know that some parrot species give their offspring unique names that are used throughout their lives. The names assigned by the parent birds differ only slightly from their own names, which makes me wonder: Is this like having a family surname, with the subtle variations being like first names?

Since then, I have been trying to decipher the specific sounds that flock members associate with one another, and I think I have figured out a few, particularly with chicken friendships. For example, my Bantam Buff Brahmas, Panko and Storey, are pretty much inseparable. I've noticed that when they do happen to be apart, they will call to each other using specific sounds that I have not heard in any other context. One or the other will call "Ur" or "Ur-ur," and the other one perks up, drops whatever she is doing, and runs to her friend.

I can only surmise that these calls are their unique names/nicknames or a sound the two of them use to give information to each other, or perhaps it means something like, "Hey, friend!" Or is it that they recognize each other's unique voice? Interestingly, the rest of the flock does not pay attention or respond to these calls between Panko and Storey.

Different Roles in the Flock

Within every flock, some chickens take on distinct roles that are often unrelated to the pecking order. Each of these roles comes with a specific responsibility, usually serving a purpose for maintaining structure and order (alpha rooster and head hen) or watching out for predators (sentinel). It is still a mystery how individual chickens fall into these roles. Are they selected by others, or do they just start taking on the responsibilities?

Many of these roles are easy to observe once you start looking for them. For example, the head hen is at the top of the pecking order. She leads the flock in their daily free-ranging activities and keeps order among the hens. The sentinel keeps an eye out for danger, frequently checking the sky and surroundings for potential threats and often perching on an elevated lookout spot for a better view. The flock heeds her alarm call, repeats it, and scatters to find shelter.

You will probably also notice that roles change as flock members age or when new hens are introduced. Here are a few more to look out for.

Explorer/investigator. This chicken tends to be braver than the others. She is often the first to try new things, taste new foods, or explore new locations. Sometimes she's even brave enough to cross the road! There is usually just one explorer/investigator, but sometimes she

If a flock has more than one rooster, one of them will take the role of alpha over all the others, known as beta roosters. The alpha rooster is not part of the hens' pecking order. He acts as the primary protector and keeper of overall flock order.

has a companion or two—accomplices, if you will. The head hen may fill this role, or it may be another member of the flock. If the explorer is not the head hen, only her close friends will follow her.

Brooders. All hens *can* go broody—meaning they develop a strong desire to sit on their eggs until they hatch—but usually only some of them do. Often it is the same two or three hens in a flock that go broody time and again. The brooders are your constant source of potential mother hens in case you decide to hatch some eggs.

Attendance taker. This chicken is often the last one to go inside the coop at night. Every evening, once everyone has gone inside to roost, the attendance taker pops back outside one more time to check for stragglers. She patrols the area quickly and then returns to the roost once she is satisfied that all are together safely inside.

Free-range monitor. Like the sentinel, this hen keeps an eye out, but in this case, instead of looking for danger, she is checking up on the flock when it's free-ranging. The free-range monitor periodically looks around and accounts for the others. If she notices that a hen is missing, she tries to spot her. If unsuccessful, she calls out to the others to signal that someone is missing. Once she makes the call—"Uh uh uh uh uh uht"—everyone else stops what they're doing and looks around for the missing chicken. When the chicken in question hears this call, she comes running back to the flock, and they all return to their activities as if nothing has happened.

Ambassadors. Some chickens want nothing to do with people, even if they have been reared as chicks and handled in the same manner as other chickens in the flock. Others genuinely enjoy interacting with humans. The ambassadors seek you out, follow you around, and want to spend time with you. These same chickens often seem to enjoy interacting with other animals as well.

Dust-bathing coordinator(s). You might notice that one or two of your chickens are the primary makers of dust-bathing holes. They dig and scratch a suitable hole, sometimes working together. When they vacate the bath, other hens will use the crater that they created.

Dust bathing serves two functions. It's primarily a way for chickens to control parasites by working dirt and dust down to their skin. But it also serves as a relaxing social ritual that a pair or even several hens can enjoy together.

When Henopause Happens

Perhaps the biggest change that you will notice about your hens is when they begin what I call henopause. A chicken's natural life span averages 5 to 8 years. Hens begin laying eggs when they are about 18 to 22 weeks old. They are at their most productive for 2 or 3 years, after which time most hens' egg production goes down. By age 5 or 6, laying becomes sporadic. Older hens often stop laying during the shorter days of winter or when replacing their feathers during a molt. Eventually they stop laying entirely.

I joke that I have both a chick nursery and a nursing home for hens. Just because hens have gone through henopause does not mean that they do not have worth or value. I believe these older ladies are the wisest. It was pretty cool to observe what they could teach the next generation, like the time I realized my first chickens had passed down my name to subsequent flocks (see page 50 for more about that).

Older chickens also teach new flock members about the neighborhood predators to be wary of and show them the best foraging spots in the yard and their favorite dust-bathing areas. They also teach the newcomers about the humans they know, such as which ones are trustworthy and kind and which ones to avoid. Some chickens also model naughty behaviors, like how to empty out the chicken feeders to find a stray mealworm that accidentally got dropped with the feed.

We can all learn from our elders, and I like to think that my senior hens are revered by the younger ones. For the girls in their golden years, the rules of the pecking order are not as strict. The other hens move aside and let the older hens eat and drink. They happily share dust-bathing spots. It appears the older hens are no longer thought of as a threat to rank but are now respected for the lessons and knowledge that they willingly share with the others.

Older hens still have a place in the flock, even if their roles change with age.

Reordering the Ranks

Similar to a queen bee, the head hen remains at the top of the pecking order until her death, unless she becomes too ill to maintain her authority. Other flock roles change as the hens grow older. Hens will keep their roles until the age of 5 or 6—about when they reach henopause—and then the torch is passed to a younger, sprier chicken. I have witnessed this with the sentinel position.

The transition happens without much fanfare. There doesn't seem to be any competition, squabbling, or jockeying for the sentinel position. I might notice the new sentinel standing taller and perching higher while the current hen still fills the role. Then one day I realize that the new sentinel is perched in the "official sentinel spot." In my coop, that's on the top rung of a certain ladder.

We all know that one poor chicken lives at the bottom of the pecking order. Even though she is part of the flock, she is last to eat and seems to get the most warning pecks from the others to back off. She is usually agile and quick and alert, always keeping an eye out for someone above her to assert their superior rank. It must be exhausting! But whenever new chicks are added to the flock, the pecking order is shuffled. I usually add chicks when my oldest hens are around 6 years old. Every single time, the hen at the bottom of the original pecking order moves up. A new girl takes her place at the bottom, and the rest of the positions are reestablished.

3

FROM FEATHERS TO FEET

HOW CHICKENS EXPERIENCE THE WORLD

Reading a chicken's mind starts with understanding how her body works. Although chickens are anatomically like humans in some ways, obviously many things set them apart from us. They lay eggs, are covered in feathers, and have wings and scaly legs and feet. Chickens also have certain capabilities that we do not—some of them downright enviable.

For example, when an injury such as head trauma or a stroke causes some of our brain cells or nerve tissue to die, the damage cannot be reversed. Instead, human brains discover different viable nerve pathways that already exist, circumventing the damaged area. Some lasting deficits might remain, and, if so, a full recovery isn't possible. Chickens, in contrast, can regenerate nerve cells and tissue, giving them the ability to completely restore brain function after a traumatic brain injury.

ON THE OUTSIDE

Ears

- Chickens have internal ears, which are covered and protected by tufts of feathers that do not impair hearing; the tufts are thought to assist with the detection of vibrations.
- Like humans, chickens have an outer, middle, and inner ear as well as a tympanic membrane (eardrum) and pharyngotympanic tubes (a.k.a. eustachian tubes) that connect the inner ears to the roof of the mouth. In birds that fly, these tubes function to adjust air pressure during flight to prevent damage to the eardrum.

- Information from the fluid-filled semicircular canals in the inner ear, combined with information from the eyes and proprioception input from the body, helps chickens regulate balance while walking and during their short bursts of flight.
- Chicken brains interpret sounds more quickly than humans do.

Outer Ear

Nostrils

- The nostrils are located at the base of the beak. A magnetic compass near the nostrils aids in navigation.

Eyes

- In addition to an upper and lower lid, chickens have a nictitating membrane, an inner eyelid that protects and helps clean the eye.
- Chickens have better eyesight than humans during daylight. They cannot see in the dark because they have very few of the rod cells that aid in night vision, but they can see daylight 45 minutes before humans can.

- Chickens are tetrachromatic, meaning they can see red, blue, and green light as well as ultraviolet. Humans cannot see ultraviolet light.
- Chickens have a 300-degree range of vision, compared with a human's 180-degree span.

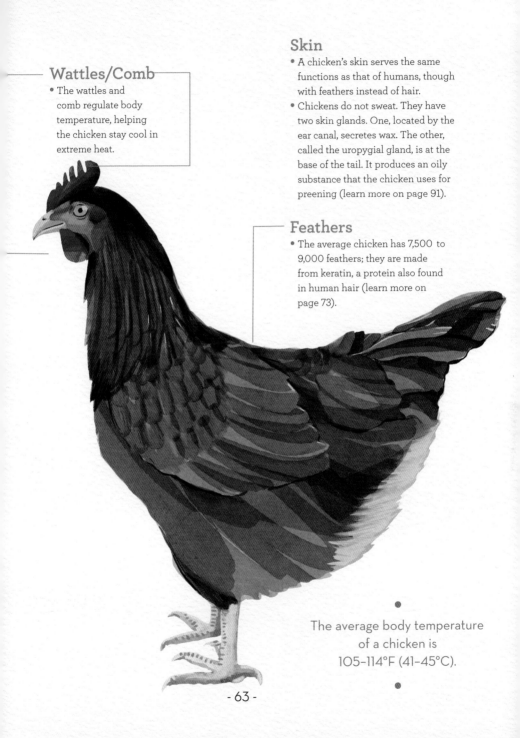

Wattles/Comb

- The wattles and comb regulate body temperature, helping the chicken stay cool in extreme heat.

Skin

- A chicken's skin serves the same functions as that of humans, though with feathers instead of hair.
- Chickens do not sweat. They have two skin glands. One, located by the ear canal, secretes wax. The other, called the uropygial gland, is at the base of the tail. It produces an oily substance that the chicken uses for preening (learn more on page 91).

Feathers

- The average chicken has 7,500 to 9,000 feathers; they are made from keratin, a protein also found in human hair (learn more on page 73).

The average body temperature of a chicken is 105–114°F (41–45°C).

SKELETAL SYSTEM

Bones
• Chickens have 120 bones. Hollow bones mean lower body weight, which aids in flying.

Nails
• The toenails are made from keratin, the same protein found in the feathers.

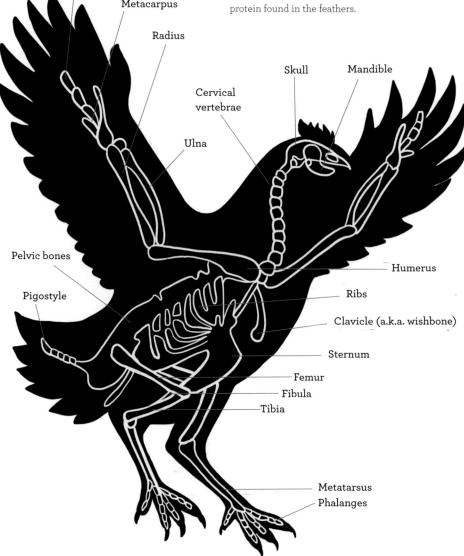

Phalanges

Metacarpus

Radius

Skull

Mandible

Cervical vertebrae

Ulna

Pelvic bones

Pigostyle

Humerus

Ribs

Clavicle (a.k.a. wishbone)

Sternum

Femur

Fibula

Tibia

Metatarsus

Phalanges

Some Chicken Facts & Figures

✦ The average life span of a backyard chicken is 5 to 7 years, but they can live up to 15 years.

✦ All chicks are born with black eyes. Their adult eye color develops by 4 or 5 weeks of age.

✦ The most common eye color in chickens is brown, ranging from amber to reddish-brown. Other colors are black, brown, buff, and white. Some chickens have speckled eyes or eyes of two different colors.

✦ Earlobes can be white, red, or blue. Earlobe color sometimes indicates the color of eggs a hen lays. A chicken with white earlobes typically lays white eggs. Red earlobes correlate with brown or dark-colored eggs.

✦ Chickens produce saliva but have few taste buds and thus a limited ability to taste. They cannot detect sweetness.

✦ A day-old chick weighs about 1.4 ounces (40 grams). Most laying hens weigh between 5 and 13 pounds (2.2–5.9 kilograms). Roosters typically weigh between 10 and 18 pounds (4.5–8 kilograms)

✦ A chicken's brain is about the size of a marble.

ON THE INSIDE

Lungs

- Unlike people, chickens have rigid lungs that are fixed in place. Chicken lungs have nine air sacs: four in each lobe and a shared one in between. The rear lung air sacs can store oxygen even if the chicken is not breathing.
- Because the air sacs in the lungs connect to the skull, clavicle, keel, humerus, pelvic girdle, and vertebrae, a broken pneumatic bone can cause difficulty breathing.
- Tiny hairs line the trachea to move trapped debris from the respiratory system. Scavenging cells within the lungs also help remove debris and kill bad bacteria.
- Because of their complex respiratory systems, chickens are extremely sensitive to aerosol particles (dust, feather dust, manure dust) and gases (carbon dioxide and ammonia from poultry waste).
- Chickens breathe 30 times per minute; people 18–20. In excessive heat, a chicken's respiration rate can climb to 150 times per minute.

Ovaries

- Both roosters and hens have internal reproductive organs but no external ones. Like all birds, hens have two ovaries, but only one is functional, usually the left. This is thought to be an adaptation to lighten body weight for ease of flying.

Kidneys

- Chickens have kidneys but no bladder. They do not produce urine. Instead, their excreta is a combination of feces and urate (that's the white cap on top of a dropping).

Crop

- Food is stored here before proceeding through the digestive tract. When empty, the crop sends hunger signals to the chicken's brain.

Heart

- The heart has four chambers and is larger in relation to its body size compared with that of humans. A chicken's heart beats around 400 times per minute, about four times faster than a human's.

Stomach

- The stomach or proventriculus has two parts. Food mixes with digestive enzymes in the proventriculus, then it moves to the gizzard, where mechanical digestion occurs.

Small Intestine

- The small intestine comprises the duodenum and lower small intestine. Food is completely digested in the duodenum before passing to the small intestine, where nutrients are absorbed.

Ceca

- These pouches produce bacteria that break down undigested food.

Large Intestine

- This is where water is absorbed from ingested material.

Cloaca/Vent

- The cloaca has digestive, urinary, and reproductive functions. It's where digestive waste mixes with urates. The vent is the external opening through which waste passes and eggs are laid.

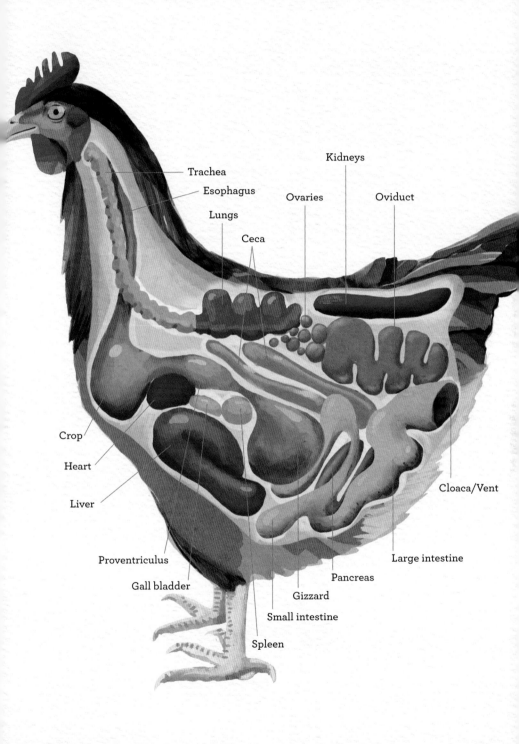

A Bit about the Beak

I've always likened a chicken's beak to the human hand—in fact, if I didn't have hands, I'd want a beak instead! The beak is a multipurpose tool, used for self-defense, preening, eating, feeding young, and investigating the surroundings. At the very end of the lower beak is a cluster of sensitive neural receptors called the bill-tip organ. This organ allows the chicken to decipher small tactile differences, much as your fingers and palm can interpret whether you are holding a stone or a marble.

The practice of debeaking, also called beak trimming—the removal of part of the upper beak, or part of both upper and lower beaks—is common in commercial poultry production. It is done to keep chickens from hurting one another in large groups, but it is nonetheless cruel. Any damage to a chicken's beak is not only physically painful but also psychologically stressful. It can affect the chicken's ability to use this highly sophisticated tool. A chicken with a damaged beak will show signs of pain, tucking the beak under a wing and minimizing its use until it heals.

The Nose Knows

✦ While baby chicks are still in their eggs, they begin to use their sense of smell as a tool to understand their environment, such as identifying their mother's scent.

✦ Chickens use their sense of smell to determine mating preferences and diversity genetics.

✦ Chickens can sense the presence of predators by their odor.

✦ Scientists have inferred that chickens use scent to assist in the recognition of other chickens familiar to them, such as those in their flock.

The Power of the Chicken's Ear

Chickens have a highly sophisticated auditory system that surpasses a human's ability to hear. Chicken embryos begin to hear at about 12 days old, while still incubating in the egg. Chickens do not have external ears, but like all birds they have a special tuft of feathers called auriculars that protect the ear canal.

Anatomically speaking, chicken ears work much like human ears in that they have a middle ear, an inner ear, and an eardrum. They also have a cochlea in the inner ear and an auditory nerve that transmits sound to the brain, which recognizes and interprets the sound information.

Chicken use their ears to detect both low and high frequencies, including infrasound frequencies that are too low for humans to hear. Examples of natural events that produce infrasound before they happen include avalanches, earthquakes, ocean waves, meteors, and volcanoes. Some animals—whales, octopuses, pigeons, hippos, elephants, rhinos, guinea fowl, and, yes, chickens—use infrasound to communicate. Humans can't do that!

Another difference between humans and chickens is that their hearing remains sharp and intact until their death. Some types of hearing loss in people are associated with the loss of tiny little hairs, called cilia, in the inner ear. On the top of each cilium is a bundle of 30 to 300 smaller hairs called stereocilia. These hair cells are responsible for helping us to hear both soft and loud sounds.

Humans are born with approximately 16,000 cilia in each ear. As we age, the number of cilia decreases; the lost ones cannot be regenerated. Baby chicks are born with fewer cilia (around 10,000 per ear), but regeneration of lost ones is no problem. You'll never find chickens in need of hearing aids!

Hearing is the last sense that is lost during the dying process for people. I wonder if this is the same for chickens.

The Beauty and Function of Feathers

Birds are the only animals that have feathers, and they have them even if they can't fly. Feathers serve many purposes, including flight, insulation, protection, waterproofing, and camouflage. In addition, feather colors and patterns help aid many birds, including chickens, in identifying one another, and in some species help to differentiate males from females.

Some birds communicate by creating specific noises with their feathers. Birds use three types of feather communication: fluttering, wing whirring, and percussion. Researchers liken these sounds to human sounds, such as clapping our hands or stamping our feet.

Fluttering is typically observed during courtship or done by fledglings begging for food.

Wing whirring is like the whistling sound that a mourning dove can make when it takes off.

Percussion, the sound made when the wings are clapped together quickly, may mean that a bird is happy or seeking attention from others. In addition, both roosters and hens flap their wings to show dominance and to appear larger. The different uses of the wings depend on the situation and the circumstances.

Chickens use oil taken from the uropygial gland with their beaks to coat and care for their feathers.

Feathers do not have nerves, so they do not conduct pain, but they do have a blood supply. Nerves and receptors in the skin allow chickens to adjust the position of the feathers in reaction to their environment. For example, a cold or sick hen will fluff up her feathers and sit in a hunched position. Both hens and roosters puff their feathers when feeling aggressive or threatened, most likely to appear larger in the eyes of their adversary.

This chicken clearly isn't feeling well, as shown by her puffed feathers and hunched posture.

Types of Feathers

Chickens have several types of feathers that serve specific functions.

Webbed feathers cover the wings and most of the body and make up the tail. They provide protection from wind and rain.

Downy feathers provide warmth when it's cold outside.

Filoplumes are hard to see. These hairlike feathers add another layer of warmth under the downy ones.

Bristles are found around a chicken's eyes, nostrils, and beak. These tiny feathers protect sensitive areas from dust and pests.

webbed   downy   filoplume   bristle

The Ayam Cemani is one of just four breeds that have black skin, meat, and bones.

Chicken Skin

Chicken skin, like human skin, contains numerous nerve endings that can sense temperature, pain, and pressure. As in humans, the skin is the largest organ in chickens and acts as a protector. Think of it as shrink-wrap, keeping everything inside.

The color of chicken skin can be influenced by diet, environment, and genetics. Free-ranging and grass-fed chickens typically have skin that is more yellow, as it contains more carotenoids than the skin of chickens that are maintained on a diet of grain.

A few chicken breeds are notable for having black skin—their meat, organs, and bones are also black. This condition, called fibromelanosis, is a genetic mutation that causes excessive pigmentation. My favorite Silkies are among this group, along with the Ayam Cemani of Indonesia, the H'Mong of Vietnam, and Svart Hona of Sweden.

The Silkie's black skin and bones are prized in traditional Chinese medicine for possessing properties that are reputed to treat disease, prevent aging, build muscle mass, and aid in recovery after childbirth.

Chicken Legs and Feet

All vertebrates (living organisms with a spine) have skin covered with scales, feathers, or fur to provide protection and warmth. Birds are unique in the animal kingdom for having both feathers and scales. The scaly, skinlike surface of the legs is called the podotheca.

Chickens have three types of scales, like those found on reptiles, on their legs and feet.

Reticulate scales are located on the footpad. They are made of alpha-keratin.

Scutate scales are large, rectangular scales distributed top to bottom on top of the toes and on the rear talon.

Scutella scales, smaller than scutate scales but also rectangular, are distributed in a side-to-side fashion on the sides of the toes. Scutate and scutella scales are made from two types of keratin: the inner surface from alpha-keratin and the outer surface from beta-keratin.

Chicken legs come in several colors, including white, yellow, blue, and green. When young hens start laying, you will notice that the color of their legs fades to a more subdued shade. This is because the same substances that give the egg yolks their bright yellow coloring (xanthophylls) also color the hen's legs. Legs can grow quite pale in heavy egg layers that are unable to keep up with replenishing their xanthophyll stores.

Most breeds have four toes; a few have five. Breeds with five toes are the Silkie, Dorking, Faverolle, Houdan, and Sultan.

Foot feathering is a genetic trait seen in only a handful of breeds, including the Brahma (shown here), the Faverolle, and the Pekin, a bantam breed. Chicken breeds with feathered feet are known as ptilopods.

Spurs

All chickens have spur buds on their legs. In roosters, the buds may grow to be an inch or two long, ending in a sharp talon. Less commonly, hens also develop spurs as they grow older. Hens of certain breeds, such as Polish, Rhode Island Red, Australorp, and Orpington, are more likely than others to have spurs. Spurs are used for fighting; roosters often use them, but hens rarely do.

Chickens experience REM sleep, the place in the sleep cycle where dreaming happens. I think you must have some imagination to dream, and I like to wonder what chickens dream about.

4

CHICKEN PSYCHOLOGY

EXPLORING THE REALM OF BIRD BRAINS

Chickens live active and full lives. Whenever individuals of any species live in a socialized group setting, there are interesting dynamics and interactions. Despite their marble-size brains, chickens are complex and sophisticated beings.

Unlocking the mind of the chicken has led to some surprising and unexpected discoveries: They have good memories. They are capable of deceitful and manipulative behavior. They experience a variety of emotions and form strong attachments. There's a lot going on in those bird brains!

How Do Chickens Learn?

Animals of all species use their senses to process information about their environment. For humans, information processing related to objects in our physical environment is sequential and systematic. Typically we observe our surroundings with our eyes while simultaneously using our auditory and olfactory senses to gather more information. We then move on to exploring with our hands. This type of experiential learning helps us gather information and make decisions. For example, we taste a new

food and decide that we like it, so we seek it out in the future; if we don't like it, we avoid it when we come across it again. Chickens do all of this, too, except they use their beaks instead of their hands.

We also learn by observing others. This is social learning. A timid child watching a daredevil kid clambering to the top of the jungle gym can judge that this activity is fun and exciting but also potentially harmful. Watching the daredevil succeed or fail can help the timid child decide whether to climb onto the jungle gym or stay put. Chickens do this, too. As they watch other chickens, they gather information to make decisions: How does a new toy in the run work? Is that new food worth trying? How do I feel about the household's new puppy? Is it safe to venture inside to explore the house and nibble on the cat food?

Chickens also use observation to solve problems and get what they want. I once accidentally dropped some scratch into the chicken feeder on top of the regular feed, instead of scattering it on the ground as usual. One smart chicken quickly emptied the entire feeder searching for every morsel of the tasty scratch. The other hens watching her realized the possibility of that particular feeder containing scratch. After that one incident, every single chicken would work to empty that feeder out as quickly as possible, but they ignored the feed once it was on the ground. I eventually outsmarted them by making a guard for the feeder out of chicken wire so they wouldn't keep wasting feed. And I was much more careful about tossing scratch farther away from the feeders!

Social Learning in Social Circumstances

Chickens, like people, use social learning when determining if something is worth facing a risk. It comes into play all the time in establishing and maintaining the all-important flock hierarchy. When a new hen is introduced to an existing flock, the lower-ranked hens closely observe her interactions with the higher-ranking hens.

Let's say a hen named Rosy sits in the middle of the pecking order. A new hen, Penny, is added to the flock. Instead of challenging Penny herself, Rosy watches to see how Penny will react when challenged by a hen—let's call her Clara—who is above Rosy in rank. If Penny loses the challenge, Rosy might decide that she can take on the newcomer and improve her own rank. But if Penny wins the encounter and becomes dominant over Clara, Rosy can infer that she also would lose to the new hen and thus learns to avoid Penny.

Chicken Tricks

Chickens are quite trainable and can learn many different behaviors, as is readily seen in many online videos. Clicker training, which pairs a reward sound with a treat, is very effective, as most chickens will do almost anything for something tasty. Chickens can be taught to recognize different shapes, colors, and pictures and to identify them by pecking at them.

They can learn to complete complex obstacle courses, waddling up and down ramps, weaving through poles, and even jumping through hoops. A piano-playing chicken once appeared on the TV show *America's Got Talent*. Training chickens is fun, rewarding, and joyful for all involved.

Passing Down Knowledge

Like humans, young chickens learn from the older members of their family. For example, a mother hen teaches her chicks where to find water and what foods taste good and are safe to eat. In one experiment, researchers fed a group of chickens two types of colored corn, blue and yellow. The blue corn was treated with a chemical that made the chickens feel ill.

The chickens quickly learned to avoid the blue corn. Later, when one of those hens was raising chicks, blue and yellow corn was scattered around their living space. The mother steered her chicks away from the blue corn, indicating that she remembered it wasn't good.

By watching how their mother interacts with flockmates, chicks also learn about the hierarchy of the flock and their place in it. For the first three months of their lives, chicks are under their mother's care, and the flock regards them as an extension of her. Thus, the chicks share her rank in the pecking order. They learn from her how they must behave in relation to the other members of the flock. However, once the chicks begin to assimilate into the existing pecking order between 12 and 14 weeks of age, they must jockey for their own position. Some of them may come out above their own mother!

What's My Name?

I have witnessed an amazing instance of knowledge being passed along by my dear hen Tilly from my original flock. As I've mentioned before (see page 50), she named me with a particular sound that I came to recognize because it was unlike any other sounds I heard from the flock: "Bup, bup, bup, baaahhh." She started it, and the others in the flock picked it up.

I wasn't sure if Tilly's flockmates would continue to use my name after Tilly passed. You can imagine my surprise and delight when I went to the coop after she was gone and heard the flock still calling out my name when they saw me. To me, it meant I was still part of their lives.

Chickens use visual and olfactory cues to recognize as many as 100 different flockmates.

Getting to Know You

By this point, it should come as no surprise that chickens are capable of higher brain function than most people think. In the past couple of decades, great strides have been made in understanding how intelligent chickens are and how they experience the world. Chickens make conscious decisions about social relationships, forming deeper bonds with some flockmates than with others. Researchers have discovered that chickens recognize up to 100 different members of their flock, using visual information such as the size and shape of combs, and feather pattern and color.

It appears that they also use smell as a tool for recognition—specifically, the scent of the oily substance secreted from the uropygial gland used for preening the feathers. For many bird species, this oil is not just for cleaning and waterproofing. Its complex chemical composition helps birds identify individuals, which is an important part of maintaining a diverse gene pool. Birds prefer to mate with unrelated individuals and can distinguish by scent which prospective mates are the best choice.

Members of the same breed are often found spending time together within a flock—perhaps scent is one way in which they recognize their similarity. After witnessing this tendency in my own flocks and hearing about it from other chicken keepers, I always encourage people to get at least two hens of the same breed in their mixed flocks.

The sebaceous material from the uropygial gland contains vitamin D precursors, which convert to the active form of vitamin D when exposed to ultraviolet light during preening. So, in addition to waterproofing the feathers, chickens incidentally ingest the vitamin D, which combines with calcium to keep their bones strong and is deposited in the egg yolk.

"Hawk Attack!"

Another sign of their intelligence is that chickens use physical and vocal signals to exchange information. This is called referential communication and has been observed in many species of animals. For animals like chickens who are always alert to the danger of predators, being able to communicate about a threat is an important behavior. Chickens use different calls to identify different predators. When they hear a certain alarm call, the flock reacts by running for cover to hide from an aerial predator; a different call will have them looking around to see where a ground-based threat, like a fox, is coming from. (See also Chickens Can Understand Different Species, page 32.)

In one study at Macquarie University in Sydney, Australia, roosters were shown videos of predators. As they watched the videos, the roosters made different alarm calls based on the type of predator in the video. For example, the rooster gave a call specific to hawks. Further studies revealed that a rooster will change not only the volume of his call when he sees hawks overhead but also the length of his call based on his location when alerting the flock. Roosters are known to increase their calls when they perceive they are hidden safely within the landscape.

Deceit and Manipulation

While researching this book, I was surprised to learn that chickens are capable of devious behavior. Indeed, there can be a dark side to these birds, and it involves something dubbed Machiavellian intelligence. First coined in 1997, the term describes the use of deceptive or manipulative techniques to take advantage of others. It is a well-known type of intelligence that is thought to serve a key role in species evolution. Examples of Machiavellian intelligence noted in chickens have to do with mating attempts. It turns out that roosters have mastered the art of deception.

Roosters tidbit with vocalizations and dancing to attract mates (see page 47). Hens understand a rooster's intentions from either the vocal or physical cues. Fake tidbitting occurs when a rooster mimics the sounds and dance steps associated with tidbitting while only pretending to have a tasty morsel to share with a hen, should she go near him. When the hen approaches

Lower-ranking roosters may act out tidbitting without vocalizing, to avoid the notice of the alpha.

the rooster, she discovers that the only thing waiting for her is a rooster who attempts to mate with her. However, hens are fast learners. After being fooled a few times by the same rooster's fake tidbitting, the hens learn to ignore any rooster who is making empty promises!

In flocks with multiple roosters, there is one dominant rooster (the alpha), with all other roosters (betas) subordinate to him. Most of the time, beta roosters successfully tidbit silently, just by dancing and offering food. This is because when the beta roosters court with vocalization, they are more likely to be attacked by the alpha rooster. However, when the alpha rooster is distracted, the beta rooster will add back in the vocalizations with the tidbitting in an attempt to convince the hens that he is now the alpha rooster.

Hen Zen

People can enter a meditative state, and I wonder if it's possible for chickens as well. If chickens are capable of meditation, then brooding is when they do it. It is truly strange to visit a broody hen sitting quietly on her nest. She lies flattened over her eggs like a pancake, awake but staring blankly into the distance without making eye contact. Sometimes she neglects to eat and drink. She remains in this state much of the day, breaking only to talk to her eggs and rotate them under her brood patch or to make her big, once-a-day broody poop. But if you disturb her, she will break from her trance to growl and hiss at you, often pecking at your hands so that you will leave her eggs alone.

I'm not sure if we will ever know if chickens are capable of meditation, but it appears to me that broody hens are doing something quite similar during the 21 days that they sit on their nest of eggs.

Anticipation

The concept of time includes the ability to sense when things will happen in one's routine and to anticipate activities based on the time of day. At our old house, there was a fox den near our yard; for a while, the vixen paid the chickens a daily visit, usually between 1:00 and 3:00 p.m. When this first started, the chickens were a complete mess! They squawked and carried on every time, even though they were safe in their run. But after a while the girls learned and remembered. I noticed that around 1:00, the sentinel hen became more alert, while everyone else also seemed to have their guard up. It was as if they were waiting for the fox to appear.

The moment they spotted her, the girls would alert one another and march inside the coop. Eventually, the fox realized that a chicken dinner was not to be had and stopped coming by. I was amazed these chickens recognized the fox's schedule and did a bit of problem-solving on their own.

My favorite moment of seeing their anticipation is during our morning ritual. When they hear me calling to let them know I'm coming and then hear the clank of the metal lid on the can of scratch as I scoop some out in the nearby shed, they start calling back to me from the coop. As I walk across the yard, there they are, leaping with joy, their eyes on me and that container of scratch. After I scatter their morning treat in the run, I love to watch them cluck and coo contentedly as they scurry around finding bits of grain. While I tidy the coop, they come to check up on me and have a little chicken chat and cuddle. It's the part of my day that I anticipate the most, too!

Episodic Memory

This type of memory is the ability to recall past events that influence present and future behavior. It has been described as a sort of mental time travel. For example, if you have a delicious candy apple at the county fair one year, you are likely to remember where to find the booth that sells them when you return to the fair the next year. You might also remember going on a ride after eating that candy apple and feeling sick afterward, so you know not to do that again!

Similarly, chickens remember good foraging spots and seek out those places. They remember places where they had an unpleasant experience, such as encountering a predator. If attacked in a favorite free-ranging area, chickens will avoid that spot, at least temporarily, until they perceive it to be safe again.

I have witnessed this in my own flock. I have an unfriendly neighbor who doesn't care for the chickens. Making himself big and scary, he will yell and chase after them to shoo them out of his yard. (You should have seen his reaction when three abandoned guinea hens decided to live in the trees above my chicken coop and roam the neighborhood for months!) Not only do my hens recognize my neighbor, but they don't dare go into his yard, knowing that their presence is unwelcome.

I'm fascinated by the way a chicken's memory works. I wonder if they remember their lives as we do, in a sort of autobiographical way. Do they remember the trip from the hatchery to their new home? Do they think back on their days in the brooder? Do they think about flockmates who have passed away?

Where Did It Go?

Chicken intelligence includes the concept of object permanence. This means that when an object is hidden from them, chickens still know it exists. (Human infants develop the skill of recognizing object permanence at around 8 months.) The degrees of understanding object permanence range from level 1 to level 6. In level 1, when an object is hidden from view, the test subject thinks the item no longer exists. In level 6, the subject can follow the location of a hidden object from place to place and infer where it has landed. A classic example is the carnival shell game of moving cups, where you keep your eye on the ball hidden under the cup.

A study by researchers from the University of Italy attempted to decipher what level of object permanence chickens understand. To date, researchers have validated that chickens can understand the concept of the existence of a partially hidden object (level 3), which indicates that they recognize an object from its parts. In addition, chickens might be able to understand the existence of a hidden object (level 4).

Are Chickens Sentient?

One of the biggest debates over the last 100 years or so has been whether animals are sentient beings—that is, whether they have feelings and emotions. This is important because sentient creatures deserve basic rights and standards of care. For a long time people refused to admit that animals could feel pain, let alone experience emotions such as joy, sadness, distress, and fear. Sentience refers to a certain depth of awareness animals have about themselves and others. Primates and many other animals, including dogs, dolphins, and rats, have been proven to have levels of self-awareness, and research shows we can say the same for chickens.

To prove self-awareness, one must ask: Are chickens capable of thinking of themselves as individuals? Do they compare themselves to other chickens? Self-worth, value, and a sense of individuality all come from self-awareness, and when it comes to chickens, self-awareness is complex and sequential. A couple of studies have sought to investigate self-awareness in chickens through measuring self-control and self-assessment. Both can be linked to self-awareness.

Self-Control

We can define *self-control* as the ability to make a decision in the present that will elicit a desirable outcome in the future. For people, self-control develops around the age of four.

Chickens have demonstrated the ability to delay gratification to obtain a "jackpot" food reward. To get the jackpot, chickens must ignore access to smaller amounts of food while waiting for a larger payoff. When it comes to this experience, the chicken must incorporate memory, the concept of time, and the ability to consider different options and different results to make the best choice. From this type of real-time decision-making, researchers have inferred that chickens can make methodical and sequential choices to their benefit.

Throughout the winter, I add enrichment treat toys to the coop to keep the flock amused when they can't go free-ranging. One is a yellow plastic ball with holes that I usually fill with sunflower seeds. Another is a clear plastic container with holes that sits on a flexible pole in the ground. This I fill with mealworms that fly out when the chickens peck at it. I usually refill the treat balls at the same time every few days, starting with the sunflower seeds and then moving to mealworms, the flock's favorite treat. First, I toss out a handful of scratch to distract the girls while I work. The smartest hens ignore the scratch, and they also hold off on the yellow ball of sunflower seeds. They stand there waiting anxiously for me to fill up the mealworm container. Then I get out of the way, because they cannot wait to get their fill of mealworms before the others realize what's happening!

Chickens are capable of ignoring ordinary treats if they think higher-value ones will be offered later.

Self-Assessment

Self-assessment is the ability to compare oneself to others and identify oneself as a separate entity. Chickens do this when they figure out where they fit in the pecking order. Experiments have proved that they are able to determine the social hierarchy by witnessing interactions between those that are members of their existing flock and those that are not. The observing chicken can form logical conclusions about the pecking order and about others that help to lead their decision-making.

Simply sorting themselves and placing themselves within the flock's pecking order shows self-assessment. In flocks with multiple roosters, jockeying for the position of alpha rooster entails constant self-assessment and a high level of self-confidence, which is another component of self-awareness. For many hens, knowing their place within the pecking order and having a particular job gives them a similar sense of self-confidence.

I wonder if chickens distinguish right from wrong. We know that they follow certain rules within the flock with expectations of behavior based on rank. But do they make moral choices outside of the pecking order?

Personality and Temperament

Personalities develop like a soup of traits ordered from an à la carte menu. Everyone has a unique combination and degree of characteristic that makes them special. Some data shows that we are born with some aspects of our personality, while personality also develops from input from our environment and our interactions with other individuals. Memories, ongoing self-assessment, and learning from experiences also play a role.

Lori Marino is a neuroscientist and an expert in animal behavior and intelligence. I love her definition of personality as "a set of traits that differ across individuals that remains consistent over time." (I discuss more of her research on page 112.) Personality is what makes us individuals. In 1949 the psychologist Donald Fiske and his colleagues described five basic personality traits in people, a model that still holds up in psychology studies. These characteristics are openness to experience, conscientiousness, extraversion, agreeableness, and neuroticism.

"Personalities"—that is, behaviors that indicate temperament—are also found throughout the animal kingdom. Studies have established a similar list of five characteristics of animal temperament: exploratory tendencies, activity, boldness, sociability, and aggressiveness. Temperament refers to the innate traits that cause us to behave or react in certain ways. For example, a naturally anxious or emotionally reactive animal will respond to stimuli differently than one with a calmer or more outgoing temperament.

Some chickens are more likely than their flockmates to explore new areas.

I like to think of the way in which we are like chickens, and some of these categories of human personality traits and animal temperaments overlap in interesting ways. In some cases, I think we can use the same synonyms among species. Here are some things I've observed about some of these temperament characteristics in my own flock and the ways in which I think our feathered friends resemble us.

Exploratory tendencies. This trait matches "openness to experience" in people, though in general, chickens tend to be fearful of new things. However, they are open to exploring and investigating when led by a brave chicken. A few years ago we moved our entire chicken coop and run to our new home. Being in their familiar coop and run made it an easier transition for the girls, who enjoyed having fresh new dirt in their run to scratch in. But when it came time to free-range, they were initially hesitant and fearful. I had to accompany them and lure them out of the run with mealworms.

Eventually Tilly, always the explorer, decided it was safe, and each day she ventured a little bit farther from home. It didn't take long for the others to follow her example, and soon everyone was comfortable with their new surroundings.

Activity. I admit that finding a correlation between this temperament characteristic and the personality trait of conscientiousness in people is a bit of a stretch, but this category for people covers impulse control and goal-oriented behaviors. These behaviors have been well studied in animals, such as the experiment in which chickens learned to wait for a food jackpot (see page 101).

An example of goal-oriented behaviors is when in my own flock, my chickens figured out how to circumvent a temporary chicken-wire enclosure around the garden. One by one, they plopped themselves onto the flimsy fencing, finally crushing it low enough for them all to gain entry. Chicken teamwork and ingenuity allowed them an unsupervised afternoon in the vegetable garden. Another time, when I forgot to toss out their morning scratch before I let them out to free-range, the girls took matters into their own hands, heading right for the open shed where they knew the treats were stored.

Boldness. This trait is similar to extroversion in humans, a category that includes talkativeness, assertiveness, and a high amount of emotional expressiveness. My now departed

A sociable chicken will come running to interact with people she knows.

rooster Chocolate comes to mind when it comes to boldness. Even though he was a tiny Silkie Bantam, he was mighty. He was a sweetheart to me and surprisingly popular with the hens. Weighing in at barely 3 pounds, he did not fear predators. Once he even came to protect me from my male neighbor. As we stood chatting, Chocolate ran to me and started making a big kerfuffle in the leaves at my feet, scratching and chattering as if to tell my neighbor, "Hey, this is my girl, too, so you better be nice to her!" We both had to laugh!

Sociability. This category correlates with agreeableness in humans and includes affection, kindness, altruism, and trust. Chickens are, of course, highly social, with a strong sense of community and the ability to regulate their society through a clearly understood pecking order. They form friendships, show affection toward one another, share treats, snuggle in a dust bath together, or preen one another. Flock members look out for the group, keeping an eye on wanderers and calling them back if they go too far.

Along with affection, I have witnessed sadness and even heartbreak in my hens when flock members passed away. I once had two Buff Orpingtons who were inseparable. When Oyster Cracker passed away from ovarian cancer, her sister, Sunshine, became deeply depressed. For a while she paced in the run, calling and looking for her sister, but eventually she spent most of her time sleeping on a log. She wouldn't interact with the other chickens and didn't even want to partake in pecking for scratch. I took her to the vet, as I feared she was ill. The conclusion was that nothing ailed her other than a broken heart. She passed within a few months of losing her friend.

Aggression. In our society we discourage aggressive behavior and rely on social norms to regulate behaviors of those around us. Animals have different norms, some of which require aggression for survival. Aggressiveness can be both genetic and a learned behavior and can be demonstrated by both roosters and hens. Aggression in hens most often comes in the form of pecking, although sparring can sometimes occur. Aggressive pecking appears mean-spirited, as it is sometimes done for no obvious reason. Sadly, some chickens just seem to be born mean.

Along with the pecking will come the pulling of feathers from others. An aggressive chicken will pluck feathers from the top of other chickens' heads or the back of their necks—two places that chickens cannot reach on their own bodies. Sometimes, pecking can lead to open wounds. At this point the bullying bird should be removed from the flock.

A rooster's aggression involves not only pecking but also the use of his spurs. He will attack anyone or anything that he perceives as a threat to his flock; that includes a fellow rooster that might consider mating with one of his girls. Roosters are very possessive and may even fight to the death. Aggression and bravery go hand in hand, however. Roosters will also protect their hens with their very lives. They often usher their girls to safety, then return to take on the threat.

Fighting cocks are bred to be aggressive.

Chickens with topknots
are sometimes targeted
by other chickens
who peck at their
extravagant feathers.

Thinking about Synonyms

I remember a school assignment where the teacher asked us to come up with synonyms for the word *mean—evil, devious, selfish, nasty, malicious, cruel, unpleasant,* and so on. Even though these words have similar meanings, they convey different notions. In thinking about how people describe chickens, I've noticed that some terms used for breed characteristics make the chickens sound like humans: *friendly, gentle, inquisitive.* Other terms make them sound more like livestock: *docile, flighty, noisy.*

To celebrate chicken/human similarities, I'd change those last three adjectives to *calm, lively,* and *talkative.* Why can't we just describe chickens in people terms? Chickens can be shy or outgoing, loving or mean, serious or silly. Some are wonderful mothers; others are awful bullies. That sounds like people to me!

Feelings and Emotions

I feel completely comfortable discussing chicken emotions. In fact, it's one of my favorite chicken-related things to talk about, probably because it is the topic that surprises people the most. I, too, was surprised by my discoveries, which is what prompted me to begin writing my blog, *Tilly's Nest*, in 2010. So many wonderful things were unfurling before my eyes that I had to jot them down.

For many centuries, and even until quite recently, most people thought that animals were incapable of reasoning, incapable of having any type of language, and lacking in emotions. Scientists were more concerned about not anthropomorphizing their subjects than exploring their true capabilities. We can thank Charles Darwin for doing pioneering research on animal emotions. Working predominantly with dogs, cats, and apes, Darwin focused on body language and facial expressions as indicators of specific emotions.

Since most researchers focused on mammals in their studies on animal emotions, it took longer to accept that birds are also capable of emotions and self-awareness. Writing in the journal *Animal Cognition*, neuroscientist Lori Marino noted that, historically, scientists stopped short of using the word *emotion* in relation to birds, instead choosing *affect*, which refers to the observable manifestations (expressions, gestures, vocalizations, and so on) related to an internal emotion. Marino suspects this word was chosen in an effort not to overestimate the capabilities of animals.

I would have to agree. It is much easier to exploit chickens when we do not think of them as capable of emotions. On the following pages, I discuss some of the emotions that chickens have demonstrated to the satisfaction of researchers (ones that chicken lovers have recognized for a long time!).

Chickens are clearly capable of feeling and expressing a range of emotions.

Empathy

Not only do chickens experience feelings and emotions, but they can also sense them in others. All species have similar wants. These include being cared for, receiving empathy from others, and trust, safety, and love.

One study designed to detect empathy in mother hens put three groups, each a hen with chicks, in different situations. One group was put in a quiet enclosure, another was in a pen where they were disturbed by noise, and the third group was in a pen where puffs of air were randomly blown on the chicks and the mother hen. In that setup, the mother hen was prevented from intervening and aiding her chicks. In the first and second groups, it was motherhood as usual. But in the third group, when the air puffs disturbed the chicks, the mother became upset as well, even if she didn't experience the disturbance herself. The outside physical interaction caused her to react by squawking loudly and acting frantic.

In a follow-up study, the researchers investigated what would happen if the mother hen in the third situation could intervene. They found that when she was able to place herself between the chicks and the puffs of air, both the chicks' and the mother's levels of distress decreased.

Chickens are known to experience emotional contagion—the tendency to feel the emotions of those around them. When looking at objective indicators, such as heart rate variability and hormonal fluctuations as well as body temperatures, researchers have shown that when a few members of a flock are stressed, frightened, or suffering, other members of the flock show signs of stress as well, even if they are not experiencing the same stimuli. Emotional contagion serves a purpose by bolstering the flock's safety from predators and helping to enhance group living.

It's not unusual for a pair of chickens to form a close bond with each other.

The Comfort of Chickens

I believe that chickens experience empathy. This was never so evident to me as when my stepfather passed away unexpectedly. One morning after his passing, I reluctantly bundled up to do my chicken chores. I had left them to other family members for a few weeks, but I missed my girls.

As soon as I heard them calling my name ("Bup, bup, bup, baaahhh!"), my heart skipped a beat with joy. They were so happy to see me! I plopped down to sit with them. I gave them some mealworms and admired their new feathers, fresh from the molt. Long after the mealworms were gone, they huddled around me, hanging out and making their wonderful, comforting chicken sounds. Maybe they were happy just to spend time with me, and maybe they could sense that I was grieving. I'll never know.

I think the key to dealing with the pain of grief is trying to find a new normal while still allowing yourself to grieve. Surrounded by my flock, my face wet with tears, I came to realize that the hole in my heart would be filled with new memories. Life would go on, just differently. Acknowledging my feelings and my profound grief was a good thing. My chicken family helped me begin the healing process. It turns out I needed them as much as they needed me.

Pride

Proud birds hold their head high and puff out their chest, as if to boast. I think a mother hen seems proud when she first brings her new baby chicks out to meet the rest of the flock. Roosters will strut around looking proud as they spend time with their favorite hen or after they win a fight with another rooster.

But perhaps the proudest moment that I have witnessed in my flock is after a hen lays an egg. Every time, she seems so pleased with her accomplishment, giving the egg song ("Buh-gaw-gawk, buh-gaw-gawk, buh-gaw-gawk!") with such excitement that every-one joins in, cackling with delight and sharing in her pride. Talk about emotional contagion!

Fear and Stress

In her 1964 book, *Animal Machines: The New Factory Farming Industry*, the animal welfare activist Ruth Harrison addressed intensive animal farming, coming up with five freedoms that she believed were essential for safeguarding animals that are cared for by people. They are:

1. Freedom from hunger and thirst
2. Freedom from discomfort
3. Freedom from pain, injury, or disease
4. Freedom to express normal behavior
5. Freedom from fear and distress

Fear is a necessary emotion that arises when danger is perceived; it helps us react appropriately. It is a universal emotion across mammal and avian species, as it is crucial to survival. However, prolonged fear in chickens can lead to stress, suffering, illness, and decreased production (weight loss in meat birds and fewer eggs in layers).

Chickens experience stress from many sources, including feeling threatened by predators, being bullied, living in overcrowded conditions, and suffering illness or injury. When experiencing heat stress, chickens will pant with their mouths wide open and will hold their wings away from their sides. Chickens experiencing difficult conditions can also spike a fever or lose feathers and undergo stress molts, which can occur any time of the year.

As mentioned previously, most of the initial research on chickens arose from industrial poultry farms looking to reduce

Chickens who have plenty of room and outlets for boredom are less likely to start feather picking and other undesirable behaviors.

stress to increase productivity for both meat and eggs. The studies themselves have often included acts of cruelty: administering shocks, injecting chickens with a toxin that makes them ill, flicking their heads, and subjecting them to foul tastes. Fear in chickens is typically measured by an increase in heart rate and the bird's attempts to avoid or escape a negative stimulus.

Frustration

Chickens show frustration when certain needs are not met, such as when they are kept in living spaces that are too small and when they are bored. Frustrated hens will pace back and forth. Their calls will become longer. You will hear the gakel call (see page 41). Boredom often leads to behaviors such as feather picking and eating their own eggs or those of other hens. These undesirable learned behaviors are difficult to curtail once they begin in the flock. Maintaining a well-kept, tidy coop and run with adequate space and access to quality feed and fresh water— as well as places for roosting, perching, dustbathing, and chicken activities—helps to prevent your flock from experiencing these feelings.

Happiness

For many decades, animal welfare organizations have questioned society's prevailing attitude that animals, especially livestock, don't experience happiness. Pet owners know that dogs and cats are capable of many emotions, and if we're observant, we can tell how they are feeling. For example, research has shown that a happy dog wags its tail to the right, while a left-wagging tail indicates nervousness or fear.

Happiness, or at least contentment, has been heavily researched in chickens because happy birds are more productive. Researchers from the University of Guelph in Canada attempted to measure chicken happiness by testing 16 different breeds, looking at their physical and behavioral responses to things like

Happy chickens are excited to see you, and they run to find out what you might have brought for them. They are bright and engaged with their environment. They cluck and coo contentedly as they dart to and fro, exploring and enjoying the day.

fake worms, food barriers, and enrichment to their living space. These studies showed that birds who received mental stimulation and access to the outdoors and green spaces were happier and as a result laid more eggs and developed into better-quality meat birds. Studies like this have helped to improve the quality of life of factory-farmed birds, but we still have a long way to go. Factory farming focuses primarily on the bottom dollar and not quality of life. It is a business, after all.

Do Chickens Play?

Play behavior is considered an attribute of higher intelligence. It is well known that many animals play—a few examples are chimpanzees, otters, and ravens. Animals do not engage in play when they are scared or injured, or if food is scarce.

Some poultry researchers believe that chicken behavior such as springing into the air and frolicking could be deemed as play. I have seen chickens playing with items that deliver a reward, such as balls that deliver treats when moved around. I know they enjoy hopping on a chicken swing to rock back and forth, playing the xylophone or a toy piano to hear beautiful chicken compositions, and pecking curiously at a mirror.

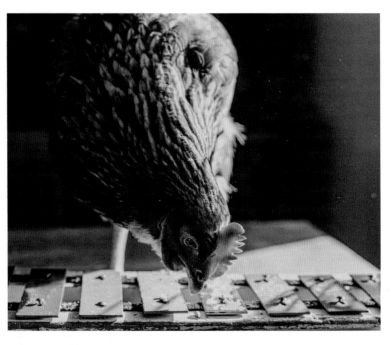

Final Thoughts

I want to touch on one last question people often ask me: Do chickens have souls? This question is probably the most controversial in the world of animals, and the debate will continue. So let me leave you with this: The word *soul* is defined as "emotional or intellectual energy, the spiritual or immortal part of the animal or human." It is said that this part survives the body after death.

According to Aristotle, the soul is divided into three parts. The nutritive soul is responsible for a being's physical nutrition and growth. It could be considered the drive to keep living. The rational soul is responsible for insight and judgment. It is the part that makes decisions based on the environment; this is where logic lives. The third part is the appetitive soul. This part is responsible for desires and wants. Some say that the soul is what forms our unique personalities; it distinguishes us and makes each one of us special.

So, what do you think after learning a bit about these wonderful birds? Do chickens have souls? I enthusiastically answer yes! After spending so many years with chickens, I am sure that they do indeed have souls. It has been an honor to know and love each and every one of them. They have been part of my family, and they continue to teach and remind me every single day what life is all about.

> Animals are a window to the soul and a doorway to your spiritual destiny. If you let them into your life and allow them to teach you, you will be better for it.
>
> **—KIM SHOTOLA**
> *The Soul Watchers: Animals' Quest to Awaken Humanity*

ACKNOWLEDGMENTS

First and foremost, a huge thank-you to all the researchers and scientists dedicated to helping improve and enrich the lives of chickens. I know that you understand how wonderful these birds can be. Thank you to my rock-star team at Storey Publishing. I cannot even begin to imagine writing books without you all by my side. I am so grateful to you for always being open to my ideas and helping me reach people around the world. Lisa, my editor extraordinaire, is always there for me, encouraging me and being open to my crazy-chicken-lady ideas. I'm sure that my readers will thank you, too, as you know that I write from the heart rather than the grammar book.

Of course, eternal gratitude goes to my wonderful family, who continue to love, support, and encourage me when I am up in my "nest" pecking away at the keyboard as I try to be a voice for chickens in the world of humans.

And finally, but most important, thank you to everyone who has supported me and *Tilly's Nest*. You have enriched immensely this amazing, unexpected experience. Life's path is not always straight; it can have turns and detours, valleys, mountaintops, and sometimes pauses. I'm so glad I was distracted all those years ago by the chickens I saw on the other side of the road! What a journey it has been.

FURTHER READING

Space prohibits listing all the materials I've consulted over the years. Here are a few key sources. For a comprehensive list, please visit tillysnest.com.

BOOKS

Ackerman, Jennifer. *The Genius of Birds*. Penguin Books, 2017.

Montgomery, Sy. *Birdology: Adventures with Hip Hop Parrots, Cantankerous Cassowaries, Crabby Crows, Peripatetic Pigeons, Hens, Hawks, and Hummingbirds*. Atria, 2011.

Nicol, Christine J. *The Behavioural Biology of Chickens*. CABI, 2015.

ARTICLES

Bekoff, Marc. "The World According to Intelligent and Emotional Chickens." *Psychology Today* (blog), January 3, 2017.

Butler, Ann B. & Cotterill, Rodney M. J. "Mammalian and Avian Neuroanatomy and the Question of Consciousness in Birds." *The Biological Bulletin* 211, no. 2 (October 2006): 106–27.

Edgar, J. L., et al. "Avian Maternal Response to Chick Distress." *Proceedings of the Royal Society B* 278, no. 1721 (October 22, 2011): 3129–34.

Evans, C. S. "Cracking the Code: Communication and Cognition in Birds." In *The Cognitive Animal: Empirical and Theoretical Perspectives on Animal Cognition*, edited by Bekoff, M., Allen, C., Burghardt, G. M.: 315–22. MIT Press, 2002.

Marino, Lori. "Thinking Chickens: A Review of Cognition, Emotion, and Behavior in the Domestic Chicken." *Animal Cognition* 20, no. 2 (March 2017): 127–47.

INTERIOR PHOTOGRAPHY CREDITS

© Alter-ego/Shutterstock.com, 40;
© Anatolii/stock.adobe.com, 84;
© Anderson Shelton/Shutterstock.com,
33; © Andyworks/iStock.com, 31;
© aprilphoto/iStock.com, 58;
© ArtMarie/Getty Images, 10;
© Ashley M Woods/Shutterstock.com,
72 t.r.; © Baz251286/iStock.com, 91;
© bazilfoto/iStock.com, 109; © Bill
Coster/Alamy Stock Photo, 80;
© Carla Freund/Shutterstock.com, 60;
© Caterina Trimarchi/Shutterstock.
com, 69; © Catherine Falls Commercial
/Getty Images, 21, 122; © Cavan
Images/Getty Images, 55, 87;
© DenisNata/Shutterstock.com, 99;
© Digital Vision/Getty Images, 96;
© driftlessstudio/iStock.com, 98;
© Ekaterina savylova/Getty Images,
23; © Eric Isselee/Shutterstock.com,
70; © Featherly/Shutterstock.com,
115; © feri ferdinan/iStock.com, 13;
© FlamingPumpkin/iStock.com, 79;
© Francesco Carta fotografo/Getty
Images, 76; © georgeclerk/iStock
.com, 52; © GlobalP/iStock.com, 1, 2,
36; © Irina Kozorog/Shutterstock.com,
74; © Itsik Marom/Alamy Stock Photo,
81; © Jamil Bin Mat Isa/Shutterstock
.com, 14; © Juniors Bildarchiv GmbH
/Alamy Stock Photo, 86; © KHON
SUPAN/Shutterstock.com, 28;
© Krzysztof Bubel/Shutterstock
.com, 30, 37, 47, 71, 78, 89;
© Liudmyla Liudmyla/iStock.com, 88;
© LUNAMARINA/iStock.com, 102;
© Majna/Shutterstock.com, 9, 45,
117; © Manhattan001/iStock.com, 56;

© Marcel Derweduwen/Shutterstock
.com, 68; © Marcos Assis/iStock
.com, 95; © Maria Kallin/Getty Images,
119; © martin-dm/iStock.com, 116;
© Microfile.org/Shutterstock.com,
110; © MVolodymyr/Shutterstock.com,
124; © N_Design/Shutterstock.com,
121; © Natalya Erofeeva/Shutterstock
.com, 90; © Nevena1987/Getty Images,
26; © Nick Beer/Shutterstock.com, 72
b.r.; © Nynke van Holten/iStock.com,
43, 77; © PAPA WOR/Shutterstock.com,
111; © Patri Sierra/Shutterstock.com,
72 t.l.; © PCHT/Shutterstock.com, 27;
© PetPics/Alamy Stock Photo, 72 b.l.;
© Popova Valeriya/Shutterstock.com,
113; © Rebecca Pope Photography
/Shutterstock.com, 54; © RyanJLane
/iStock.com, 25; © SimonSkafar
/iStock.com, 5; © SolStock/iStock
.com, 100; © Sonsedska/iStock.com,
16; © stockphoto mania/Shutterstock
.com, 19; © Terryfic3D/iStock.com,
38; © TomasSereda/iStock.com, 107;
© Tomatheart/Shutterstock.com, 35;
© Trea/stock.adobe.com, 105; © Tsyb
Oleh/Shutterstock.com, 82; © VMJones
/iStock.com, 44; © Voren1/iStock.com,
51; © Wassana Panapute/Shutterstock
.com, 93; © Wirestock/iStock.com, 123;
© Wizard Goodvin/Shutterstock.com,
46; © Zocha_K/iStock.com, 49